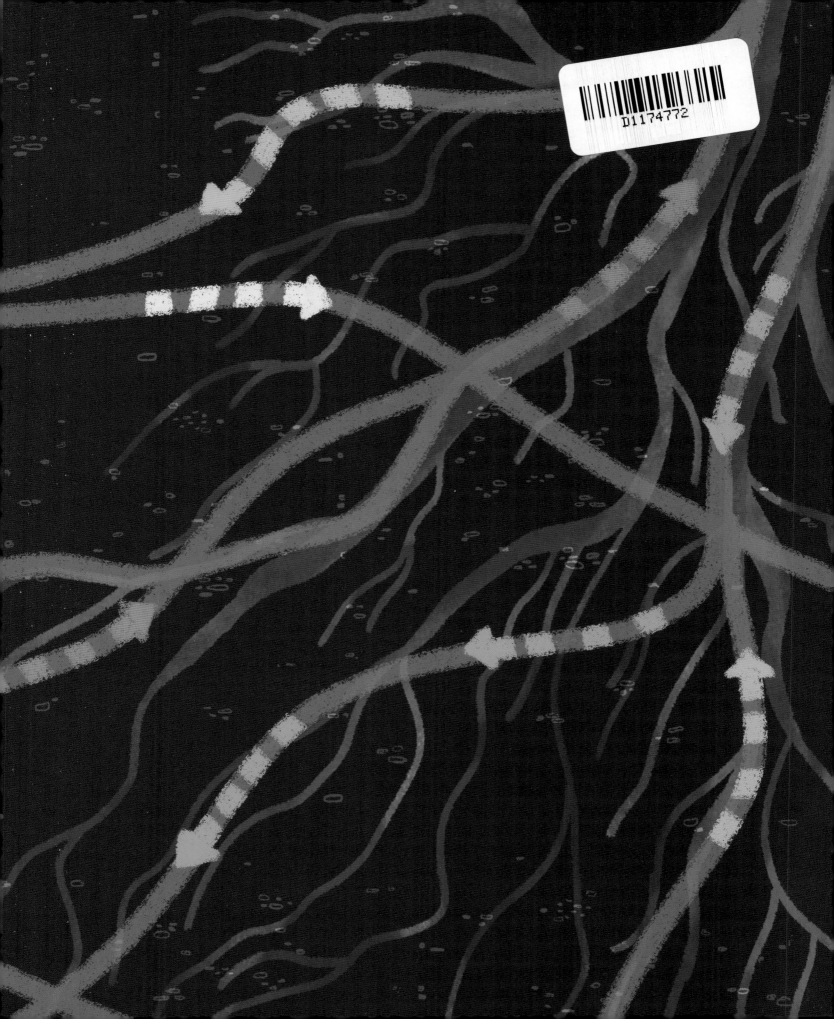

Text © 2022 by Tera Kelley
Illustrations by Marie Hermansson
Cover and internal illustrations and design © 2022 by Sourcebooks
Internal images © Shaun Cunningham/Alamy Stock Photo, Doug Wilson/Getty Images, redfishweb/Getty Images, randimal/Getty Images, fotolinchen/Getty Images, Kenneth Keifer/Shutterstock, Irina Vinnikova/Shutterstock, yarbeer/Shutterstock, Manfred Ruckszio/Shutterstock, KYTan/Shutterstock, Susie Hedberg/Shutterstock, Bildagentur Zoonar GmbH/Shutterstock
The full color artwork was created in Photoshop.
The characters and events portrayed in this book are fictitious or are used fictitiously. Any similarity to real persons, living or dead, is purely coincidental and not intended by the author.
Published by Dawn Publications, an imprint of Sourcebooks eXplore
P.O. Box 4410, Naperville, Illinois 60567–4410
(630) 961-3900
sourcebookskids.com
Library of Congress Cataloging-in-Publication Data is on file with the publisher.
Source of Production: Leo Paper, Heshan City, Guangdong Province, China
Date of Production: October 2021
Run Number: 5023477
Printed and bound in China.
LEO 10 9 8 7 6 5 4 3 2 1

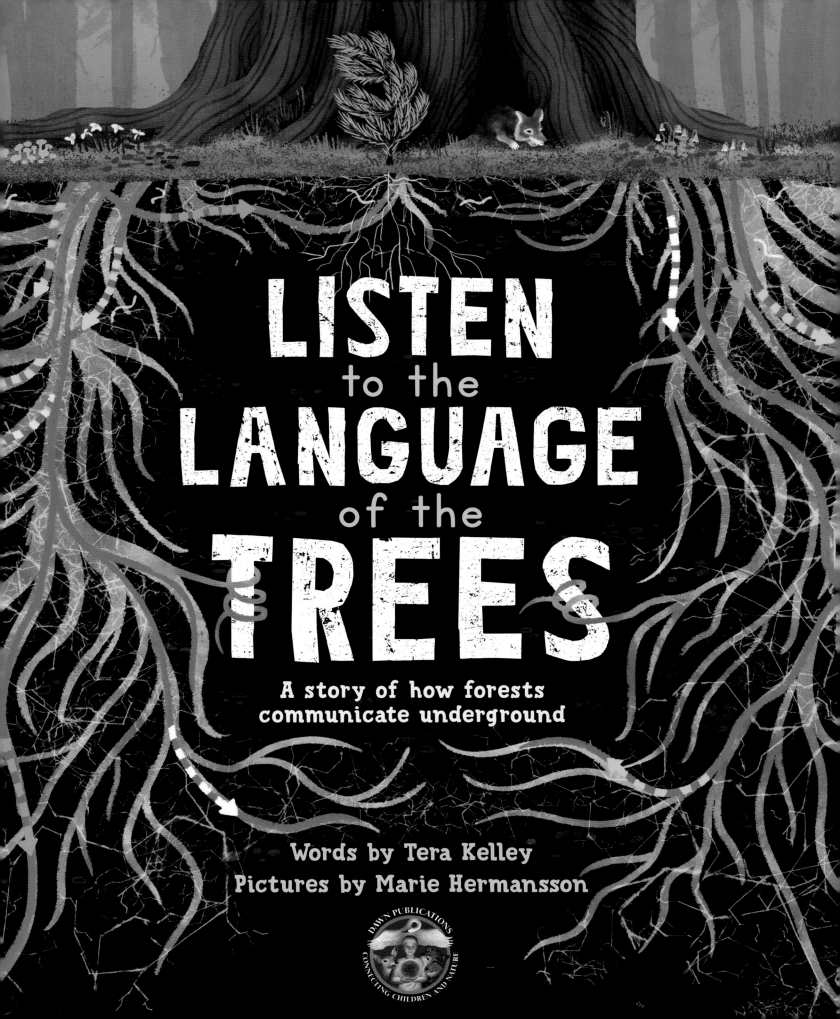

LISTEN to the LANGUAGE of the TREES

A story of how forests communicate underground

Words by Tera Kelley
Pictures by Marie Hermansson

DAWN PUBLICATIONS
CONNECTING CHILDREN AND NATURE

The forest was full of chattering. A jay shrieked at a sniffing coyote. A squirrel scolded them from a low branch. The trees rustled, as if whispering to each other.

No one noticed the tiny seedling pushing its way above the soil. Quietly, its needles stretched out in the cool darkness of the forest floor.

It was a miracle it sprouted at all.

As a seed, it had been tucked inside a seed cone hanging high on the branch of a giant tree. Till one day...

The squirrel grabbed the cone and scampered down the tree. He buried it in a special spot for a winter snack.

The seed rested underground for months. Luckily, the squirrel had a lot of special spots. He never came back!

Nestled inside its cone, the seed waited.

When it was ready, the seedling sprouted at the base of a giant tree.

Centuries ago, this mighty tree was a tiny seedling too. Now it towered over all the other trees in the forest, strong and straight. It looked like it could survive anything.

But no part of the forest is entirely free from danger.

As small as it was, the seedling's roots reached underground and met the roots of other trees. Threaded between them was a silky net of fungi, a web that stretched from root to root and beyond. Through this web, the forest passed secrets.

The trees talked to each other through their roots. They spoke of danger—of drought and pests. They spoke of what one tree needed and another had to give.

The seedling began to listen.

The giant tree with its huge crown gathered an abundance of sunlight. It soaked it up and turned it into food.

It took care to send the little seedling enough to nourish it.

How did it know the seedling was one of its own?
It was a mystery, buried under the soil where their
root tips intertwined.

It was not only the tiny seedling that depended on the giant tree, but the rest of the forest too.

Larger than all the others, the giant tree sent nutrients through the underground web. It brought water up from deep in the Earth.

It housed the owl that nested in its upper limbs. And the coyote with its den at its foot.

Day by day, the seedling inched taller.

Then, one night, everything changed.

The sky came alive with lightning flashes. The forest swayed under the force of the winds, and thunder roared in the sky. Tucked safely down

A massive strike lit up the giant tree's crown.

Thud! Down went an uppermost branch.

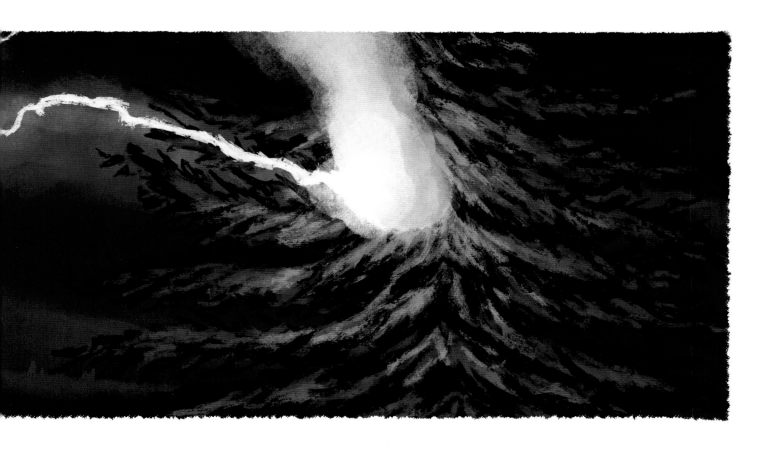

The animals of the forest scattered. The jay shrieked. The squirrel ran.

For a moment, the giant tree lit up with an orange glow. And then, darkness.

In the morning, it was quiet.

The giant tree had been struck. It had burned briefly, but it still stood tall.

Down on the forest floor, the seedling was missing one of its branches, snapped off when a limb had fallen from the giant tree. Otherwise, it had escaped harm.

But something was different.

A streak of golden light touched the seedling's needles. Instead of only drinking from its roots, it drank in the full light of the sun. For the first time, it made its own food.

After the storm, the animals returned to the forest. The squirrel began its seed cone collection again. The owl took a long nap.

The forest waited. The forest listened.

That's when the first winged beetle landed.

The beetle sent out an odor, which said, *Attack!*

Winging through the air, crawling over bark, beetles clustered on the giant tree and laid their eggs.

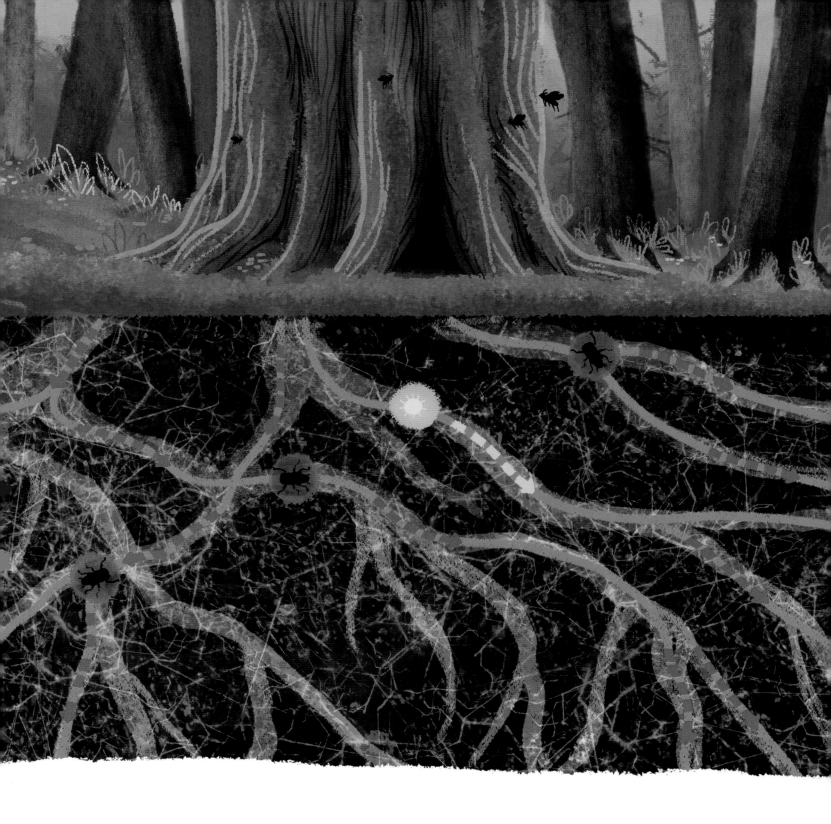

The giant tree sent out an alarm: *Danger!*

The warning pulsed through the forest, passing
through tree roots, carried by the web of fungi.

Day after day, the beetles attacked, a small but persistent swarm.

No more sugar flooded the underground web from the giant tree.
Just distress.

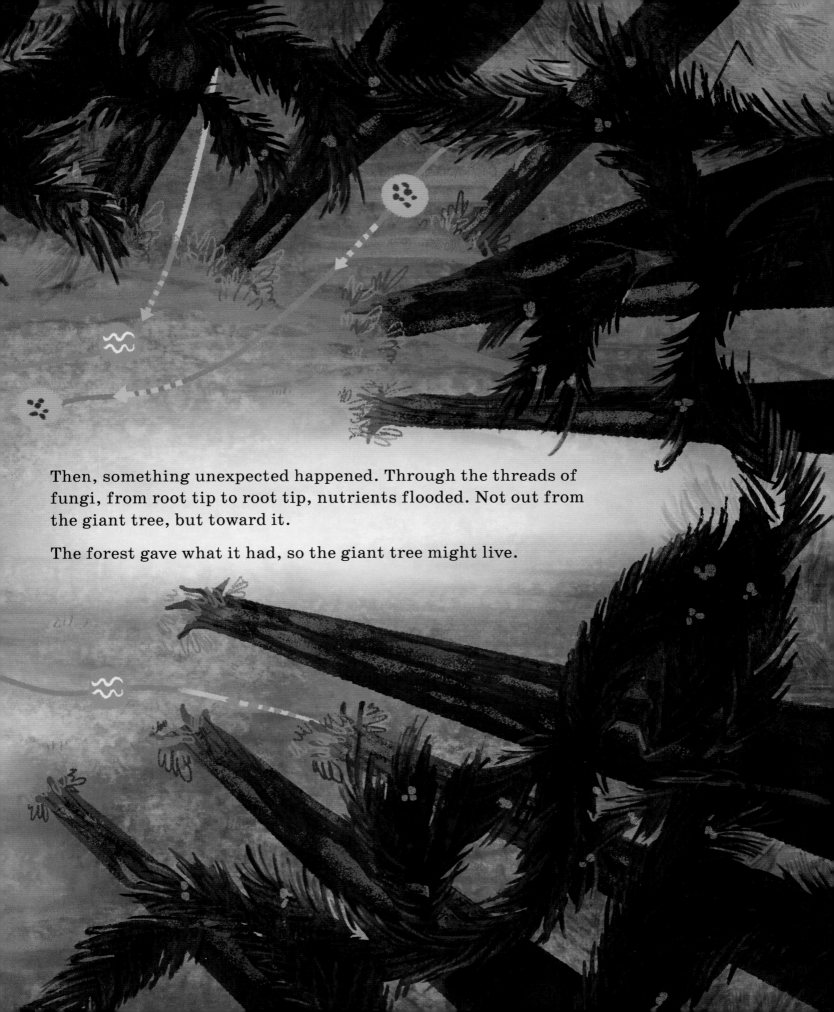

Then, something unexpected happened. Through the threads of fungi, from root tip to root tip, nutrients flooded. Not out from the giant tree, but toward it.

The forest gave what it had, so the giant tree might live.

And it did.

Someday, with luck, the seedling would grow tall too.

But for now, it drank deep through the web at its roots and waited and listened as the forest talked.

THE SCIENCE IN THE STORY

A Douglas fir tree can live to be hundreds of years old and grow as tall as an apartment building. But when it first sprouts from a seed, it is smaller than the palm of your hand.

Douglas fir trees make seeds that are hidden inside cones. These cones stay on the tree until they are ready to open. Then they release winged seeds, which are carried long distances on the wind.

Trees, like all plants, make their own food from sunlight, air, and water. This is called photosynthesis.

Squirrels sometimes eat seed cones or take them to bury and store for winter food. If the squirrel forgets to dig the cone up again, the seeds have the chance to sprout!

Have you ever seen a mushroom growing on the forest floor? These mushrooms are the above-ground part of fungi. Some fungi spread through the soil under forests. Trees grow their roots through this web.

Fungi can't make their own food like plants can. Some fungi partner with trees to get the food they need. Fungi receive the food that the trees have made with photosynthesis. In exchange, they deliver nutrients, water, and even messages between trees.

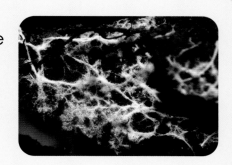

Lightning is more likely to strike the tallest tree in a forest. Some trees catch on fire and burn up from the inside when struck by lightning.

Large older trees often act as "mother trees." They might not be related to all the trees in the forest, but they do take care of them. Seedlings especially need their help.

SCIENCE CONNECTION

The trees in the story are Douglas firs, which are a coniferous (or cone-bearing) tree that grows in the western part of North America. Douglas firs spread through mountain ranges all the way from Alaska to Mexico. Douglas firs are evergreens, which means they keep their needles year round.

Douglas firs are not the only trees that partner with fungi to trade carbon, nutrients, and water. In fact, most land plants form relationships with fungi! Plants are sometimes connected to other plants through underground webs of roots and fungi. These webs are called mycorrhizal (pronounced *my-kuh-ry-zul*) networks. They are very common in forests.

Trees receive valuable help through these networks. By sharing resources, all the trees in the network stay healthier. But why do fungi partner with trees?

Fungi are more like animals than plants. They can't make their own food through photosynthesis. Fungi harvest carbon (carbohydrate sugars) from decaying material around them. But receiving carbon from trees is a lot easier than getting it for themselves!

Fungi can receive up to 30 percent of the carbon passing through the underground network. In exchange, fungi provide nutrients and water that they harvest from the soil. They also exchange messages between trees. It's a good deal for both trees and fungi!

Scientists are continuing to research how mycorrhizal networks work. There is much more to discover about these amazing underground webs!

What We Know

- Resources and information are shared through the mycorrhizal network
- Trees can recognize their kin and send them extra resources
- Seedlings that are connected to the mycorrhizal network do better than seedlings that are not connected
- Trees can emit distress signals that travel through the network
- Neighboring trees may care for sick trees by sending them resources

What We Don't Know

- Do trees control where resources are sent in the network?
- Do fungi decide which resources get sent to which tree?
- Do resources always flow from trees that have the most to trees that have the least?
- Do "mother trees" ever receive resources back from the network?
- Most research has been done in laboratories. How do these networks work in actual forests?

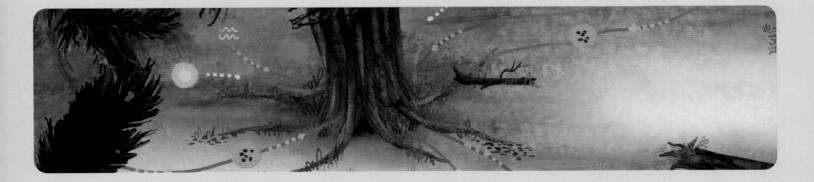

SOCIAL EMOTIONAL LEARNING

Learning about the connections and cooperation in the forest is a natural springboard for teaching about relationship skills, one of the core competencies in social and emotional learning. It includes the ability to communicate clearly, listen well, cooperate with others, resist inappropriate social pressure, negotiate conflict constructively, and seek and offer help when needed.

Activity — Kids Helping Others

The story shows the connections and cooperation in the forest through an underground network. Review the pages and ask children to find examples of other connections and cooperation, such as:

- The squirrel helped the seedling by bringing the seed cone down the tree, burying it, and then not coming back to dig it up. This helped the seedling sprout!

- The giant tree helped the seedling get nutrients and water when it first sprouted.

- The forest helped the giant tree when it was damaged by the beetles.

Explain to children that there are important connections and examples of cooperation in their life too. Ask them to think about the people who help and support them at home, at school, and in their neighborhood. These people might be parents, family members, guardians, neighbors, teachers, friends, teammates, or classmates. Then ask them how they connect and cooperate with others.

Have children fold a piece of paper in half and write two headings: *People Who Help Me* and *People I Help (or People I Want to Help)*. Ask them to list in words or pictures a person and situation under each heading. For example, *My neighbor helps me take care of my dog. I help my grandma by raking the leaves in her yard.* Conclude by reminding children that, just like in nature, our survival depends on connections and cooperation.

Activity — Kids Helping the Forest

Explain to children that just as forests help people survive, people can help forests survive. Because forests are connected in a complex web of cooperation with all other parts of nature, whatever we do to help one part of nature will ultimately help forests. It's all about connection and cooperation! Have children brainstorm possible projects. Here are a few ideas to get you started:

- Is there a park or nature area near you? Take a trip to pick up litter.

- Plant a seedling and take care of it. (Hint: It doesn't have to be a tree seedling.)

- Start a worm bed to improve the soil. You'll be helping plants get the nutrients they need.

- Recycle paper products! You can also reuse paper for taking notes or drawing.

For Richard Rodrigue, always mischievous,
curious, and kind. Your words planted the seed
for this book. I wish you were here to see it!
—TK

For Ella and Espen, may you always keep your
curiosity of the natural world.
—MH

© Bruce Malnor

Tera Kelley grew up reading among the towering Douglas firs and pungent cedars of the Pacific Northwest. As a children's bookseller for six years, she watched kids race, skip, and crawl in the direction of the children's book section with the same eagerness she felt at that age. She is currently a freelance writer and editor in Northern California, where she works at a local library. Find out more about her work at terakelley.com.

Marie Hermansson is an illustrator who strives to create illustrations that cultivate imagination and spark natural curiosity. She has a lifelong interest in plants, nature, and design, which led to a degree in landscape architecture and ultimately to illustration, her true passion. When she's not working, she enjoys gardening, hiking in the forest with her family, and going to museums. She lives with her two children and Swedish husband in North Carolina.

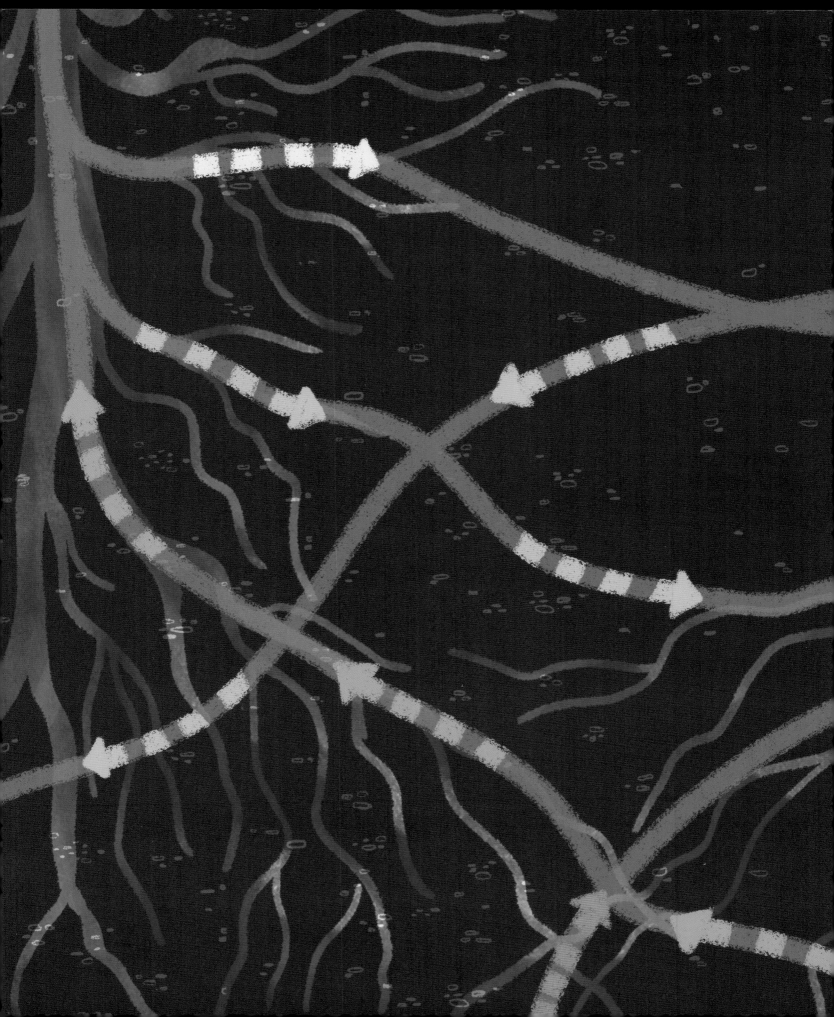